Jonathan Americo

Island connectivity

Jonathan Americo

Island connectivity

A study of the presence of Spartina alterniflora on the Cananéia and Comprida Islands

ScienciaScripts

Imprint
Any brand names and product names mentioned in this book are subject to trademark, brand or patent protection and are trademarks or registered trademarks of their respective holders. The use of brand names, product names, common names, trade names, product descriptions etc. even without a particular marking in this work is in no way to be construed to mean that such names may be regarded as unrestricted in respect of trademark and brand protection legislation and could thus be used by anyone.

Cover image: www.ingimage.com

This book is a translation from the original published under ISBN 978-3-330-76061-5.

Publisher:
Sciencia Scripts
is a trademark of
Dodo Books Indian Ocean Ltd. and OmniScriptum S.R.L publishing group

120 High Road, East Finchley, London, N2 9ED, United Kingdom
Str. Armeneasca 28/1, office 1, Chisinau MD-2012, Republic of Moldova, Europe
Printed at: see last page
ISBN: 978-620-8-34137-4

"One step forward and you're no longer in the same place... "

C.S.

Summary

Finalising this work is an act of **gratitude.** To all those who have helped me in some way. Who taught me how to organise and disorganise in the right way.

Thank you, Sueli! The nicest person I've ever met. The nicest person I could ever meet. Your help and affection are part of my life story. You generously welcomed me into academic life in São Paulo.

To my parents, who always showed me the importance of studying. Who encouraged me to seek out different forms of knowledge. Who are the example of parents in my world.

To Professors Euler Sandeville Júnior and Sérgio Tadeu Meirelles, who taught perhaps the most intense course I have ever experienced.

For the opportunity to spend time with Yara Schaeffer Novelli and her admirable knowledge.

To Luiz Bittar Venturri and Marisa de Souto Matos Fierz who welcomed me on my arrival at the University of São Paulo.

To everyone who has taught me so much, both inside and outside the disciplines. To the subjects themselves, and all that they have made me seek.

To Ana and the entire Postgraduate Secretariat of the Geography Department, who always took care to help with all the bureaucracy of an academic environment.

To my beautiful friends. To friendship, which remains easy and full, even though our encounters are so rare.

To the beautiful Bárbara, who makes love the most natural way to live.

Summary

The mangroves of the Cananéia-Iguape Lagoon System, at the southern end of the coast of the state of São Paulo, have the presence of the grass *Spartina alterniflora* on their outer fringes, which plays a fundamental role in fixing sediments that could become areas for mangrove propagation. Pioneers in the flooded mangrove region, *Spartina* holds and prepares the substrate for the seedlings of mangrove species to establish themselves. The mapping of *Spartina alterniflora* on the Cananéia and Comprida Islands, where these islands border the same lagoon channel, made it possible to identify the areas prone to its growth. The low energy points of the channel's water flow proved to be favourable for the emergence of *Spartina* banks. The continuous exchange of sediments and nutrients in a lagoon environment rich in mangroves, combined with the great capacity of *Spartina* to capture and attach itself, resulted in a study of connectivity between two islands.

Key words: *Spartina alterniflora,* mangroves, sedimentation, connectivity.

Presentation

The model of biological conservation used today in the main policies dealing with the environment is based on areas and fragments, most often modified remnants that symbolise an entire region of the past that was more extensive and continuous. Protected because of their importance for biodiversity, communities, resources, etc., these fragments have remained totally or partially isolated, which has contributed to a decline in their previously emphasised importance.

The Cananéia Estuarine-Laguar region, on the extreme southern coast of the state of São Paulo, has important characteristics for preservation that can contribute to the concept of a new conservation model: connectivity between fragments. To the south-east of the southern edge of the Paranapiacaba continuum, despite its inherent isolation as an island, the estuary provides important connectivity of matter and energy between the continent, the estuary and the sea.

Based on the foundations of *landscape ecology,* this study seeks to diagnose the entire structure of the landscape; investigate the relationships between elements within and between ecosystems (spatial units); and significant changes that have occurred over time that may have altered or influenced the area in some way. The results of this research will provide the necessary support to establish the most important and favourable routes for connectivity.

Introduction

With more than 8,000 km of coastline, Brazil has great potential for using its coastline, while at the same time it can present a great risk if this important area is not properly utilised. Knowledge of coastal dynamics encompasses various spheres of knowledge (cultural, economic, scientific, etc.), including the entire history of colonisation of our country and practically all of the Americas.

> Based on the knowledge that in all the American colonial areas, the coloniser first arrived in the new lands by sea, it is clear why the coastal areas were the first to see settlement centres. Thus, all the flows of colonisation in the New World started from centres of diffusion based on the coast, which linked the explored hinterland with the oceanic routes that reached this exploration. (Moraes, 1999, p.31).

Characterised since the period of colonisation as a country that exports raw materials, the Brazilian coastline has always been the stage for port activities that intensified as the country industrialised. According to Diegues (2001, p.108), from the end of the 1960s onwards "chemical poles, industrial centres, export corridors and ports grew in number and intensity without prior assessment of the environmental impacts resulting from them", putting the ecosystems present in the coastal zone at risk, as well as their importance and functionality for the populations that are directly linked to them.

The Brazilian coastal zone is home to a wide range of ecosystems of enormous biological, cultural and economic importance. From an economic point of view, these ecosystems are the source of income for several coastal populations that depend directly on their resources. Biologically, the ecosystems in this area of interaction between sea, land and atmosphere are of enormous value as they are habitats for countless animal and plant species.

In this respect, Diegues, a scholar of the peoples of the sea (2001, p.107) recognises that:

> Coastal ecosystems are habitats for a large part of Brazil's marine resources, as they are breeding areas, permanent or temporary refuges for countless species of fish, crustaceans and molluscs caught by industrial and artisanal fishing. They are also habitats for countless species of resident or migratory birds and important animals of the Brazilian fauna (Diegues, 2001).

The importance attributed to coastal ecosystems, be it environmental, cultural, economic, etc., has common aspects that merge, generating strategies for the use of ecosystems. Areas of great biological value also represent great economic value for the communities that live off the exploitation of these resources, or even as preserved areas that are great attractions for tourist activities that also have their economic purpose. A study is therefore needed to manage these often conflicting interests.

The wide variety of natural resources that the coastal zone possesses is put at risk by poorly planned anthropogenic activities that are often aimed solely at the economic aspect, as the population that inhabits the coastal region grows. In this way, the need for Coastal Environmental and Territorial Planning is becoming increasingly clear, enabling us to move towards a positive relationship between economic development and the sustainable use of coastal natural resources.

The study area for this research is located in the Cananéia Estuarine-Laguna Complex, which has the necessary and favourable characteristics for the development of mangroves which, in turn, have the potential for the occurrence of the grass *Spartina alterniflora, the* species chosen as the bioindicator in this study. This species plays an important ecological role in nutrient cycling, sediment retention and deposition, which favours the microfauna food chain and the expansion of the mangrove's outer fringe.

The island of Cananéia is located at a point that provides connectivity[1] with other neighbouring conservation units. The city is less than 15 kilometres south-east of Jacupiranga State Park, which together with Intervales State Park, Carlos Botelho State Park, Xitué Ecological Station, Alto Ribeira Tourist State Park (PETAR) and the Serra do Mar APA Wildlife Area, form the Paranapiacaba ecological continuum.

Objectives

0 This research project's **general objective is to** study the structure and dynamics of the mangrove ecosystem present in the Estuarine Lagoon Region of Cananéia (SP), in the interfacial area between Cananéia Island and Ilha Comprida. Based on approaches from landscape ecology, island biogeography and their related concepts, highlighting the potential and importance of the area for conservation.

1 The ability of a species to utilise the resources of an area is totally related to its ability to reach that area. What determines this ability is not only the distance, but also the biophysical nature of the route that will be used, as well as the species' behaviour. Some paths facilitate or even prevent movement between areas; others obstruct the movement of species to varying degrees. Due to the great importance of species movement, a third measure of landscape structure, called 'connectivity', is finalised. "Landscape connectivity is the degree to which the landscape facilitates or impedes movement between source areas" (Taylor et al., 1993).

Specific objectives

The main specific objective, which guides and orders the entire study, consists of **mapping the geographical distribution of the** *Spartina alterniflora* **species** in the area that is limited to the north by Vila de Pedrinhas (Ilha Comprida) to the Base of the Oceanographic Institute (Ilha de Cananéia), the southern limit. When the species is present, it occurs on the outer fringe of the mangrove at the edges of the channel, which delimits the study area to the shores of the Cananéia Sea (Figure 2).

Through the occurrence points of *Spartina alterniflora* and taking into account the biological aspects of the plant, it will be possible to **study the connectivity** between the islands, which is another specific objective.

Essential goals and steps emerge which, as they are achieved, provide the basis for understanding the dynamics and heterogeneity of the region.

They are:

- **structural characterisation**, a diagnosis of the fauna, vegetation, hydrography, soil, geomorphology and geology, and an analysis of the relationships between these elements and possible ecosystems;

- **connectivity, to** establish the possible key areas that guarantee genetic flow and real conservation.

Because it has favourable natural characteristics for the existence and maintenance of mangroves, and because mangroves are an economic source for local communities, the island of Cananéia is perfectly suited to the vision proposed by Marta Vanucci in her book *Os Manguezais e Nós*[2] , on the rational economic use of mangroves:

- "to try to get a clear picture of the roles that mangroves play in the coastal zone under different conditions, seeking an answer to the question of whether management with the intention of maximising the production of a single resource is, in the long term, more convenient than managing the ecosystem as a whole with the aim of optimising several resources;

- to establish the criteria that should be used to decide what should be done in each specific case for the rational utilisation of mangroves in areas of strongly conflicting interests".

Research context

From the first environmental conservation area created in the world to the present day, we have experienced conservation policies based on areas and fragments, most often remnants less modified by man, which symbolise an entire region supposedly close to the "original". Protected because of their importance for biodiversity, communities, resources, etc., these fragments have remained totally or partially isolated, which has often contributed to a decrease in their previously emphasised importance and in some cases isolation has also led

2 VANUCCI, M. The *Mangroves and Us.* São Paulo: University of São Paulo Press, 2002.

to undesirable changes in the supposedly untouched ecosystems.

On the other hand, research into the dynamics of fragmentation has shown that the connection between fragments is an important aspect of gene flow (Alves & Metzger, 2002; Metzger, 2003).

The study of ecological corridors and connectivity is relatively recent and its application is restricted to a few areas of the planet. In Brazil, since July 2000, the National System of Conservation Units (SNUC) has mentioned ecological corridors in Chapter IV: "The Creation, Implementation and Management of Conservation Units", Article 25:

> Conservation units, with the exception of Environmental Protection Areas and Private Natural Heritage Reserves, must have a buffer zone and, where appropriate, ecological corridors.

0 Federal Government maintains the Ecological Corridors Project, part of the efforts of the Pilot Programme for the Protection of Brazil's Tropical Forests (PPG7), with the aim of optimising the benefits offered by tropical forests, in line with the country's development goals. The Central Amazon Corridor and the Atlantic Forest Biodiversity Corridors make up the project at a national level. The state of São Paulo, with its public environmental agencies and academic units, is advancing its research and implementation of the Paranapiacaba Ecological Continuum, covering more than 120,000 hectares in the Vale do Ribeira region.

The study of connectivity (potentials and implementations) is fundamental to conservation and sustainability. This project, by proposing to understand the natural dynamics of

Cananéia, in all its complexity and heterogeneity, takes on this extremely important ecological role. Through landscape ecology, this work seeks to follow a path that is pertinent to understanding the relationships between natural elements, generating data and bases for better management and possible connectivity.

Study Area

Cananéia-Iguape system

The Cananéia-Iguape System, in the form we know it today, traces its origins back to 120,000 years ago when the mean relative sea level (NMRM) was 8 metres above the current level, reaching the foot of the Serra do Mar and covering the Pariquera-Açu Formation with transitional and marine clays (Sugio & Martin, 1978). At a second stage, the Pleistocene Regression made it possible to deposit beach crystals which, in turn, covered Pleistocene sandy deposits. At 17,000 years before present (B.P.) it is estimated that the NMRM was 110 metres lower than it is today, characterising this third moment with the erosion of the deposits of the Cananéia Formation by the rivers of the coastal plain. The fourth and final transgressive event, called the Santos Transgression or Holocene Transgression, saw the NMRM, 4 metres higher than today's, invade low-lying areas, forming the large lagoon area and depositing sandy-clay sediments rich in organic matter. In the fifth and final stage, the NMRM stabilises at the current level, forming regressive beach ridges, which confirm the presence of Ilha Comprida (Figure 1).

According to Besnard (1950a), the first scenario of the Cananéia-Lguape System presents a series of indentations bordered by the Serra do Mar and some islands. The second scenario would be characterised by tidal currents shaping the islands through processes of erosion and sedimentation. Mangrove forests would appear at a third stage, consolidating the finer sediments at the bottom of the rivers and canals. The last moment is represented by the current scenario, resulting from sedimentation and erosion carried out by tidal currents and river action (Cunha-Lignon, 2001).

Located at the southern end of the coast of the state of São Paulo, the system is made up of four main islands formed mainly of Quaternary sediments and Precambrian crystalline rocks. It is mainly drained by the Ribeira do Iguape River, as well as lagoon channels, gamboas and rivers that separate the islands of Cardoso, Cananéia, Comprida and Iguape. Of these, Iguape is the only artificial island, originating from the opening of the Valo Grande canal, which was built in the mid-19th century with the aim of facilitating navigation in the final part of the Ribeira do Iguape River (Cunha-Lignon, 2001). In 1978, the Valo Grande was closed by the São Paulo state government and reopened in 1983. When it was first opened, it was a 4.40 metre wide channel and today, due to erosion of the banks, it is over 300 metres long and holds a large part of the Ribeira River's flow (Besnard, 1950b).

The Cananéia-Iguape System connects to the ocean via three mouths: Icapara (north), Cananéia (centre) and Ararapira (south). As it is a coastal plain, the rivers and channels of the system are influenced both by ocean waters and tidal currents, and by fresh waters arriving from the mainland (Tessler & Souza, 1998).

The region has an average annual temperature of 23.8°C, with the highest average monthly temperature being 27.8°C in February and the lowest average monthly temperature being 19.8°C in July (Silva, 1989). The tide in the region is mixed, with an average height of 0.81 metres (Mesquita & Harari, 1983).

Rainfall averages 2,300 mm per year, with maximum values from January to March with a monthly average of 266.9 mm and minimum values in July and August with a monthly average of 95.3 mm (Silva, 1989).

Figura 1: Scheme of Quaternary evolution of the Cananéia-Iguape plain, according to Suguio & Martin (1978), elaborated by Bonetti Filho (1995) (In: Cunha-Lignon, 2001).

In terms of salinity, the system is currently characterised as partially mixed and highly stratified (type 2b), during the summer, autumn and spring seasons on spring tides (Bérgamo, 2000). During the winter, on spring tides, and in the autumn and spring on quadrature tides, the system is classified as partially mixed and weakly stratified (type 2a) (Bérgamo, 2000).

Geology and geomorphology: synthesis for the region

The Cananéia-Lguape Coastal Plain is largely covered by sandy sediments from the Quaternary period, made up of selected, homogeneous fine sands of continental origin at the base and marine origin at the top. Also part of the Pariquera-Açu, Cananéia and Ilha Comprida Formations, these deposits are based on the Precambrian Crystalline Embayment related to the Açungui Group (Karniol-Marquez, 2007).

The Crystalline Embayment outcrops around the bridge over the Cubatão canal and in the oceanic part of the Bom Abrigo and Cardoso Islands. There are also Cenozoic alkaline rocks that support the São João hill on Cananéia Island and the Morrete hill on Comprida Island (Spinelli, 2003). Embayment points also separate large plains formed essentially by fluvial-laguar marine deposits (Suguio & Martin, 1986).

The coastline of São Paulo and southern Rio de Janeiro can be divided into five main morphological units, according to the natural limits of the Crystal Embayment: (1) Cananéia-Lguape, (2) Itanhaém-Santos, (3) Bertioga-São Sebastião Island, (4) São Sebastião Island-Serra do Parati and (5) Ilha Grande Bay. With smooth transitions between one unit and another, their morphological differences can be explained by sedimentation and tectonic dynamics (Sugio & Martin, 1986).The Cananéia-Lguape Unit is formed by a coastal sedimentary plain of Cenozoic origin, of the marine type, with altitudes ranging from 0 to 20 metres and a slope of less than 2%, characterising the Iguape-Cananéia Coastal Plain (Ross, 1997).

130 kilometres long and 40 kilometres wide, it has a complex drainage network made up of lagoon channels, rivers and gamboas, with the Ribeira do Iguape being the main river on the plain and the one with the highest flow on the São Paulo coast (Suguio & Martin, 1986).

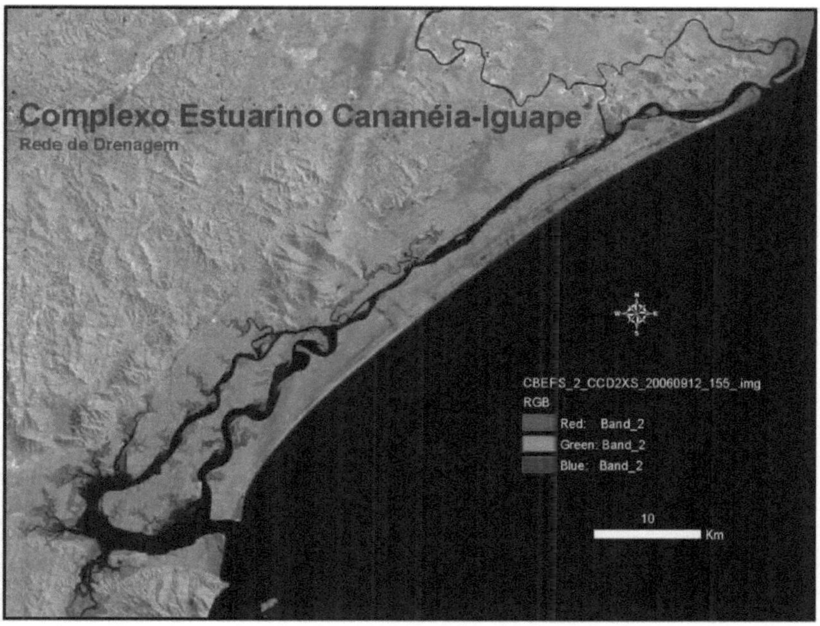

Figura 2: Composition of CBERS image (2006) highlighting bodies of water (ocean, rivers and gamboas) in black (Source/Image: Instituto Nacional de Pesquisas Espaciais, INPE).

CHAPTER 1

Spartina alterniflora

O mangroves have a unique biota. Among the herbaceous species in this study, we focus on *Spartina.* The genus *Spartina is* made up of 16[3] species of perennial grasses, belonging, according to APG II *(Angiosperm Phylogeny Group,* 2003), to the *Poaceae* family. The genus is native to the entire coast of the Atlantic Ocean, occurring mainly in North and South America. Smaller occurrences can also be found on the Atlantic coast of Europe and Africa, and also on the Pacific coast of America (Mobberley, 1956).

The leaves are characterised by being well-articulated; ciliated ligules; panicle-shaped inflorescence; spikelets with a single flower between rachis; asymmetrical, membranous glumes and, in some species, the upper glume surpasses the lemma; thin hyaline palea also surpass the lemma and completely surround the ovary and stamens; absence of lodicle; protogynous flowers; absence of seeds in most species; and asexual reproduction through scaly rhizomes (Goodman, 1958).

The species *Spartina alterniflora is* a grass adapted to estuaries, occurring in the intertidal region, where there is a variation in salinity and periods of exposure and flooding (Adaime, 1977). Amongst the representatives of the genus, *Spartina alterniflora* is the most characteristic of mangroves and is not restricted to the tidal marsh region (Klein *et a!,* 1981).

---

3 Mobberley (1956) emphasised the problem of taxonomic separation of the genus, since the criteria are vague and inconsistent, making it difficult to accurately identify some species and infraspecific taxa.

It occurs in Latin America in several countries, extending from the border between Mexico and the United States, 32°N, to Bahia Blanca in Argentina, 34°30'S (Costa & Davy, 1992). It is also distributed along the east coast of Canada and a large part of the Atlantic and Pacific coasts of the United States.

S. *alterniflora* has three distinct heights in most places where it is found: tall, intermediate and short individuals, which tend to occupy different areas in tidal regions (Teal, 1962; Adams, 1963; Squiers & Good, 1974). Tall *Spartina* is likely to occur just above the channel margin and areas that are regularly flooded. Intermediate Spartina occupies the top of the natural banks above the channel. Low Spartina occurs in the upper part of the marsh and is therefore less flooded (Teal, 1962; Squiers & Good, 1974).

Mangroves represent, in a way, a tropical and subtropical analogue of marshes, which does not exclude the possibility of also finding marshes in these regions (Costa & Davy, 1992). The distribution of *Spartina alterniflora* in mangroves seems to be closely linked to the absence or attenuation of the canopy in some part of the forest (Chapman, 1974; Reitz, 1961; Tomlinson, 1986). It is therefore restricted to pioneer formations in clearings, on the outer edges close to the channel or on the estuarine coast (West, 1977; Santos, 1989 *apud* Costa & Davy, 1992).

Studies in North Carolina (Williams and Murdoch, 1969) and Louisiana (Gosselink & Kirby, 1974) - both in the United States of America and the latter further south - have shown that at higher latitudes, the growth of 5. *alterniflora* begins in spring and ends in autumn, while at lower latitudes growth can be continuous. Thus, there is a relationship between latitude and biomass production, since this production is linked to the higher annual averages of the *standing crop* at latitudes closer to the equator (Turner, 1976). In line with this, there is also a close relationship between temperature and the growth and production of *Spartina alterniflora* banks.

In addition to latitude and temperature, other factors can influence the production of living biomass, such as tidal range, nutrient availability and salinity. As far as tidal range is concerned, there is no plausible explanation, while several studies show an increase or indifference in the possible relationship with biomass production (Nixon &Oviatt, 1973a, 1973 b; Valiela &Teal 1974).

The difficulty lies in knowing whether the tide will bring the nutrients needed to increase production. It is known that an increase in nitrogen in *Spartina alterniflora* banks leads to greater production (Tyler, 1967 *apud* Turner, 1976; Marshall, 1970 *apud* Turner, 1976; Valiela &Teal, 1974; Sullivan & Daiber, 1974; Payonk, 1975 *apudTurner*, 1976). Similarly, there is a strong correlation between ammonia and phosphorus with biomass, height and weight, while salinity does not appear to be a significant variable for these parameters. Soil water salinity has been shown to be a more influential factor than surface water salinity (Nixon & Oviatt, 1973a).

There are many factors that make up the natural dynamics of mangroves and marshes and that influence the biomass production of *Spartina alterniflora.* In addition to those already mentioned, insolation is also another important element that can alter production within the same area. The plant's efficiency in converting solar energy can vary between the edges and the highest parts of the estuarine ecosystem, which further emphasises latitude and temperature as important aspects in biomass production (Turner, 1976).

Underground biomass production accounts for a large part of S. *alterniflora*'s total biomass production. The biomass of its rhizomes and roots can be up to 2.5 times greater than the aerial biomass (Nixon & Oviatt, 1973; Cammen *et al*, 1973). Live biomass also contributes 2.5 times more than dead biomass, which indicates the plant's great capacity for vegetative reproduction through its rhizomes that produce embryos throughout the year (Panitz, 1992). *Spartina*'s phosphorus levels are high, and can reflect the plant's needs (Rodewald-Rudescu, 1974 *apud* Panitz, 1992). During the spring, the rhizomes transfer phosphorus to the small *Spartina* germs. In summer, phosphorus levels increase in the aerial part of the plant. In early autumn, when most of the individuals have died, the phosphorus returns to the plant's rhizomes. Still with a high mortality rate during the winter, there is an accumulation of detritus, which releases phosphorus into the water or sediment and which can be metabolised or released into the oxidised layers[4] . This whole process accelerates the availability of phosphorus for the biogenic circulation of the ecosystem (Esteves, 1979 *apud* Panitz, 1992).

Spartina alterniflora leaves contain a large amount of polyphenol, which makes them

4 This process was described by Esteves (1979), who called it the *Bomb Effect.*

unattractive to herbivores. The rate of polyphenol is what determines the flavour of vegetables for their potential consumers, and its concentration is inversely related to the intensity of being ingested (King & Health, 1982). Polyphenol is also important in the formation of humic substances, which is fundamental in the formation of sediments and primary production (Bianchini, 1982, 1984).

Spartina alterniflora also plays an important ecological role as a major supplier of detritus and organic energy to the ecosystem (Odum & de la Cruz, 1967; Darnell, 1967; Panitz, 1992). Organic detritus represents a storage, transport and buffering mechanism for the estuarine ecosystem: the organic material produced in a certain period is released later, either in the same place or at a point downstream; it has the capacity to make nutrients available during seasons of low primary production (Darnell, 1967b).

Organic detritus is all types of biogenic materials, in their various stages of microbial decomposition, which represent energy sources for consumer species (Darnell, 1967a). In tidal flats or in the portions of mangroves occupied by *Spartina,* organic detritus is the link between primary and secondary production, since only a small part of the production of these areas is consumed while still alive (Odum & de la Cruz, 1967). Studies carried out in Georgia estuaries have estimated that almost 95 per cent of the organic detritus on Sapelo Island is made up of *Spartina* waste, and that this organic material is produced and exported throughout the year (Odum & de la Cruz, 1967).

Around 60 per cent of the annual primary production of *Spartina alterniflora* is related to the growth of its roots and rhizomes which, once dead make this rate of carbon available to the decomposing organisms present (Wiegert, 1977). The importance of *Spartina alterniflora* for the estuary increases even more, since when it is decomposed, its detritus undergoes protein enrichment through the action of microbial populations, generating a great nutritional source for the food chain (Odum & de la Cruz, 1967).

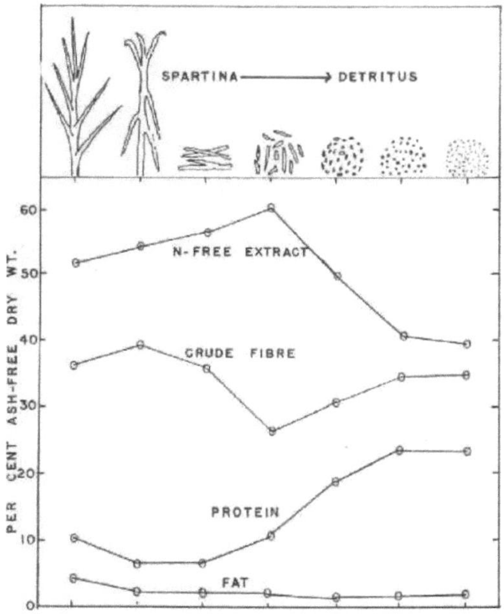

Figura 3: Nutritional composition of successive stages of *Spartina* decomposition, showing an increase in proteins and a decrease in carbohydrates (Odum & de la Cruz, 1967).

Figura 4:

Studies carried out in the salt marshes of Georgia, USA, have shown that the deposition of inorganic particles is greater, reaching a maximum in areas close to channels and open waters (less than l.5m), and gradually, but significantly, decreases as the distance inland increases (Figure 4) (Leonard *et al,* 2002).

Research on the east coast of the Delmarva Peninsula, in the state of Virgina, USA, complements the Georgia data by showing that sediment deposition processes occur during all tides, while suspended sediment concentration rates increase during high tide, which consequently increases the deposition rate (Christiansen *etal,* 2000).

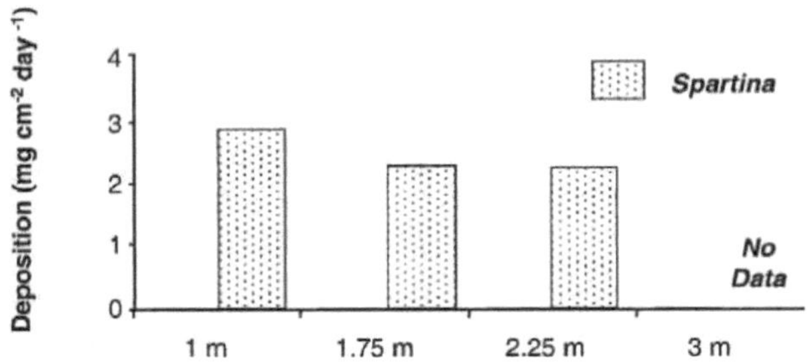

Distance from Marsh Edge

Figura 4: Average sediment deposition according to the distance from the edge to the interior of the marsh occupied by Spartina (adapted from Leonard et al, 2002).

Analyses carried out on *Spartina alterniflora* banks have indicated the inability of the tidal flow to resuspend or remobilise the sediments once deposited (Christiansen *et al,* 2000). These data, associated with its own decomposition, show the great contribution of *Spartina* to the structural formation of sediments, which can serve as a substrate for the occupation of secondary species, taking into account the perennial condition of the grass (Panitz, 1992).

The ecological role played by the *Spartina alterniflora* grass in the estuarine environment is clear, as is its importance in nutrient cycling, the supply of detritus and energy, participation in the food chain and sedimentation in the area close to the channel.

A few topics summarise the knowledge gathered in this chapter about the role played by 5. *alterniflora* in salt marshes and mangroves:

- great capacity for vegetative reproduction, and in this way give continuity to all of its role in the ecosystem;

-release of phosphorus into the water or sediments during the winter, where it will be metabolised or taken to other oxidised layers, making it available for biogenic circulation;

.a large amount of polyphenol in its leaves, which is extremely important for the formation of humic substances which, in turn, play a fundamental role in sediment formation and primary production;

-provider of detritus and organic energy, making nutrients and sediments available to the ecosystem;

-this organic detritus is the link between primary and secondary production and represents a major source of energy for consumer species;

. when decomposed by microbial populations, the detritus is enriched in protein, generating a great nutritional source for the food chain.

It can be concluded that the presence of *Spartina alterniflora* contributes significantly to the estuarine environment, both for biotic and abiotic components of this ecosystem. The number of studies carried out on this grass in its various aspects is vast, but it is worth emphasising the importance of long-term research relating *Spartina alterniflora* to mangrove environments.

CHAPTER 2

Connectivity

In order to understand the concept of connectivity, we must first understand the methodological bases that made it possible and the context in which these bases were developed. In this way, we can better understand its origins, advances and importance.

Three different approaches will be dealt with here which, in some way, have benefited the idea of connectivity, species movement capacity and the conservation of ecological processes (Benett, 2003). Each approach will be subjected to two subtopics, Arguments and Benefits, which will show in a simplified and didactic way its theoretical merits and validities. The first subtopic presents the concepts present in the approach and how they developed, and the second shows the support generated in favour of connectivity.

Natural history and wildlife management

Arguments

The observations made by field researchers have generated empirical knowledge about species and communities, which are of great importance and influence in the basic concepts of natural area planning (Bennet, 2003).

Observations of migratory movements; observations of the impacts of local barriers that influenced the flow and mortality of species; observation of the behaviour of animals and their daily or regular movements, helped to understand the importance of movement as an ecological process (Bennet, 2003).

Benefits

This empirical knowledge has benefited the planning and management of protected areas, as:

-have enabled species to cross barriers and maintain local movements;

-have helped to maintain migratory movements between different geographical areas;

-have enabled the recolonisation of habitats through increased dispersal and immigration (Bennet, 2003).

Island Biogeography

Arguments

Developed by MacArthur and Wilson (1963, 1967), the equilibrium theory in island biogeography resulted from findings that islands contained fewer species than continental areas of relative size. The theory proposed that the number of species on an island tends towards a level of equilibrium, determined by the dynamics between colonisation and extinction of resident and new species on the island (MacArthur & Wilson, 1963, 1967 *apud* Bennet, 2003). The colonisation rate is basically controlled by the island's degree of isolation from the mainland's source areas, while the extinction rate is essentially determined by the island's area.

Benefits

The analogy made between islands and fragmented areas (mountain tops, lakes, nature reserves, forest remnants, etc) has influenced the management, planning and design of protected areas. The possibility of theoretically determining a number of species in equilibrium, as well as reducing isolation and encouraging colonisation, has provided significant help for natural conservation.

The theory of island biogeography has benefited movement between fragmented areas by encouraging colonisation through the non-isolation of these areas.

Landscape ecology

Arguments

Landscape ecology assumes that every landscape, whether in its natural form or modified by man, is made up of a mosaic of different habitat types. Spatial patterns make up and influence the ecological processes present in these mosaics, which in turn change over time (Forman & Godron, 1986; Merriam, 1988; Hansson *and* Angelstam, 1991; Forman 1995; Hansson et al, 1995 *apud* Bennett, 2003).

The movements of species, water, wind, as well as the flow of energy, matter and nutrients between mosaics are essential for understanding the dynamics of the ecological functions and processes present in the landscape (Hobbs, 1993; Forman, 1995 *apud* Bennett, 2003).

Benefits

In order to understand and conserve species and communities, it is necessary to understand the capacity of these species to live within and between landscape mosaics. The movement of species between mosaics creates vectors for the flow of energy, nutrients and matter that influences the population itself and the environment (Bennet, 2003).

First uses

The first studies to work on the idea of connectivity came from analyses and experiences based on the theory of island biogeography. Data showing the relationship and influence of the size of an island, its isolation, immigration and colonisation rates on the number of species, led some authors to consider and recommend connectivity.

Preston (1962) writes about the limitations of preservation areas, showing, through his canonical distribution studies, the inability to have a smaller-scale replica of the fauna and flora of a much larger area. And he emphasises the importance of continuous corridors between preserved areas as the only remedy to prevent the isolation of the area. Even so, there would be an impoverishment of biodiversity on a national scale due to the reduction in natural areas.

Willis (1974), when studying the bird population on the island of Barro Colorado, Panama, concluded that the preservation of natural areas would be more efficient if they were not isolated or surrounded by anthropogenic uses, but rather if they were linked by "corridor zones" (Figure 5).

Diamond (1975), on the design of protected areas, emphasised the importance of connecting several separate reserves through bands of protected habitats, since these connectors can significantly increase the conservation objective. Species in protected areas would not have to cross vast areas of unsuitable habitat, especially in the case of sedentary species with restricted habitat preferences.

Connectivity

Taylor *et al.* (1993) arrived at the concept of connectivity that is most widely accepted and used in the natural sciences. Analysing the work of Dunning *et al.* (1992) on the ecological processes present in a landscape, Taylor complemented it by showing the importance of movement as a component of landscape structure (Figures 6 and 7). The ability of a species to utilise the resources of an area is totally related to its ability to reach that area. What determines this ability is not only the distance, but also the biophysical nature of the route that will be used, as well as the species' behaviour. Some paths facilitate or even prevent movement between areas; others obstruct the movement of species to varying degrees. Due to the great importance of species movement, a third measure of landscape structure, called 'connectivity', is finalised.

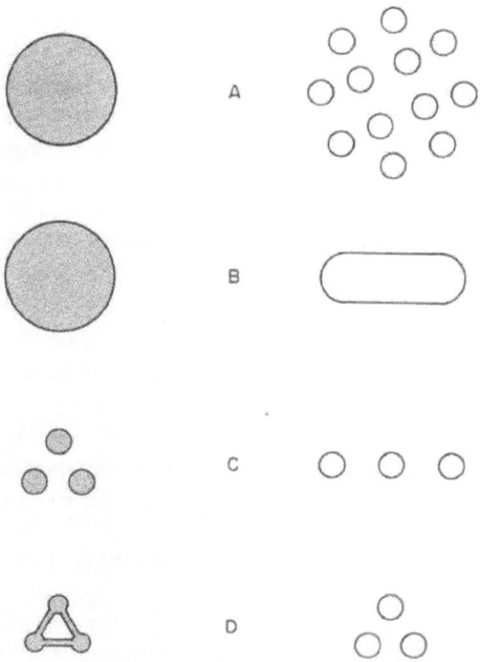

Figure 5: Geometric rules for the design of nature reserves, based on current biogeographical theory. The drawings on the left result, in each case, in a lower rate of spontaneous extinction than those on the right. Both the figures on the left and right have the same total area and represent reserves of homogeneous environments. A: a continuous reserve is better than a fragmented one, due to the effect of area and distance. B: a circular design is better, due to the edge effect (MacArthur & Wilson, 1967, pp. 115-116 and figure 37). C: aggregated fragments are better than those arranged linearly, due to the distance effect. D: if the reserve needs to be divided, extinction will be lower when the fragments are connected by natural habitat corridors, regardless of the width of the corridors (Willis, 1974). Another principle, not incorporated into this scheme, is that, regardless of the reserve design, the extinction rate can be reduced if similar reserves are close together .[5]

"Landscape connectivity is the degree to which the landscape facilitates or impedes movement between source areas" (Taylor et al., 1993).

5 Extract from Wilson & Willis (1975) Applied Biogeography. In: Cody, M. L. & Diamond, J. M. (ed) Ecology and evolution of communities. Belknap Press of Harvard University Press: Cambridge, Massachusetts and London: England.

Forman (1995) used connectivity as a measure for the connections between corridors, networks and matrices. The lower the number of discontinuities, the greater the structural connectivity, which does not define or guarantee behavioural connectivity.

These two components, structure and behaviour, define the connectivity potential for an area, species, community or ecological process (Bennett, 1990 ***apud*** Bennett, 2003).

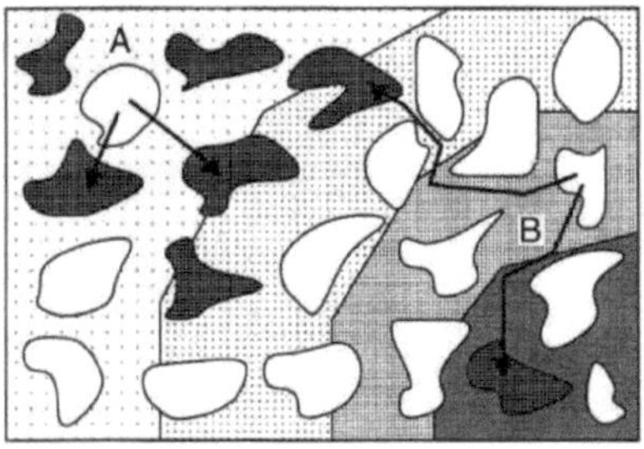

Figure 6: Landscape complementation. Landscape connectivity is based on a mosaic of fragments. Light and dark areas have different types of resources for animals. Darker areas have greater connectivity. Animals in area B can access fragments more easily (arrows), thus complementing their resource needs (Taylor et al., 1993).

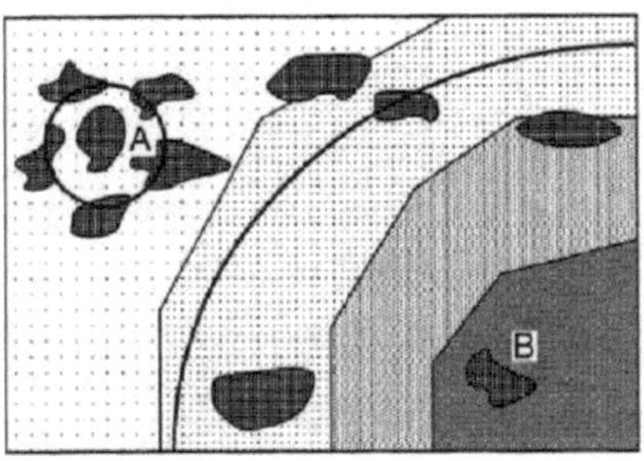

Figure 7: Landscape supplementation. Landscape connectivity is based on a mosaic of fragments. The darker areas have greater connectivity. Animals in area B can access other patches more easily than in area A (areas inside the circle), thus supplementing their resource needs (Taylor et al., 1993).

0 structural component of connectivity is determined by the spatial arrangement of the various types of habitat present in the landscape. It is closely related to the continuity of suitable habitats, the size and quantity of gaps between fragments, the distance to be travelled by species, and the presence or absence of alternative paths (Bennett, 1990 *apud*

Bennett, 2003).

The behavioural component of connectivity is related to species' reactions to the physical environment in which they live. It is influenced by the spatial and temporal scale at which movement occurs, its perception of the landscape, its need for resources, its tolerance of transformed habitats, the stage of life the species is in, and its response to predators and competitors. In this way, species that inhabit the same area of the landscape can experience different levels of connectivity (Bennett, 2003).

Analysing these two components opens up a discussion about what connectivity really is in practical terms. The scope of understanding of connectivity is broadened: corridors, in their various forms, which have long been put to use, are no longer the only means of achieving connectivity. It is not the aim of this discussion to underestimate the values and successes of ecological corridors, but rather to broaden the way in which connectivity is understood and worked with, as well as to instil knowledge.

Connections and connectors

Different types of corridors exist and are created, both by nature and by man. Watercourses, animal trails, the top of a slope, roads, power lines, canals, all of them are, with their differences, forms of connecting corridors.

Corridors are strips that differ from their surroundings and permeate the landscape. Vegetation corridors fulfil important functions and serve different purposes (Bennett, 2003):

(a) corridors protect the biodiversity of riparian vegetation, rare and threatened species, and dispersal routes; (b) corridors help water management, flood and sedimentation control, reservoir capacity, help maintain fish and fisheries; (c) linear strips can collaborate in agroforestry production by acting as windbreaks, controlling soil erosion, providing timber and preventing desertification; (d) corridors can serve as recreational and sports areas; (e) green belts can act as a cohesive factor between communities and cultures, creating identity and cultural diversity; (f) corridors are dispersal routes for isolated species in preservation areas and strips of coastal vegetation.

Corridors can differ in terms of their origin, requiring appropriate management according to their purpose (Forman & Godron, 1986):

Natural corridors, generally following the characteristics of the landscape, follow the paths of the topography, as well as the watercourses associated with riparian vegetation

Remnant corridors are the result of alteration or disturbance of their surroundings, and can be represented by forests along roads, areas not used by logging activities, or any other process that leaves linear remnants;

Restored corridors, the result of restoring previously degraded areas to their natural state;

Planted corridors, such as living fences, windbreaks and belts, generally made up of non-native species, with an obvious anthropogenic purpose;

Disturbance corridors, generated through the construction of roads, power and transmission lines, etc., presenting a line of disturbance that differs from its surroundings.

Some types of corridors have emerged in parallel with agricultural activities, used as a protection area, border delineation, as a stockpile or expulsion area for herbivores, among others.

Living fences[6] are an example of linear connectivity that occur mainly in rural areas. They vary in structure, floristic composition and origin (Bennett, 2003). They are generally of anthropogenic origin and are mostly made up of shrubs and small trees, but can also contain large trees.

Living fences act as connectors between natural and semi-natural areas in rural environments. They are strongly associated with agriculture and their composition and structure are closely related to adjacent plantations or pastures (Forman and Baudry, 1984). Green belts[7] and windbreaks[8] are also linear strips that act similarly to living fences. Also created by man, they can fulfil different functions and achieve different objectives: wind protection, reducing erosion, source of wood, habitats and aesthetics (Forman & Baudry, 1984).

The riparian zone is formed by the vegetation adjacent to watercourses (riparian forest),

---

6 Free translation adopted for "Hedgerows".
7 From the English "Shelterbelts".
8 "Windshields.

which has different structural and floristic characteristics to its surroundings. Riparian vegetation stands out as a rich habitat for fauna (Figure 8) (Murray & Stauffer, 1995 apud Bennett, 2003). This wealth of habitats is partly due to the contact interface between terrestrial and aquatic environments, which is subject to flooding, different water levels, isolated pools and successional patterns of vegetation, which contributes to a diversity of habitats and opportunities for fauna (Bennett, 2003). This factor is also associated with the excellent fertility of the alluvial soils, which favours a greater volume and structural diversity of vegetation, and the possibility of ecotones between riparian vegetation and vegetation on higher ground.

The riparian zone also plays an important role in hydrological regulation, filtering sediment and nutrients, stabilising banks, regulating water temperature and, consequently, the productivity of the system (Bennett, 2003).

Roads, created to transport and integrate a country, represent a large linear system which, with the presence of lateral vegetation, can be great wildlife reserves. In England, it is estimated that roadside vegetation amounts to a total of 212,000 ha of reserves (Way, 1977), and in the United States this figure could reach 8.1 million hectares (Adams & Geis, 1983). The vegetation along roads has great potential for structural connectivity if it is well established and managed.

Fragments of preserved areas close together can act as connectors through a sequence of movements between areas. Stepping stones are suitable for increasing connectivity in already altered landscapes, especially for species that are tolerant and able to move between fragmented areas.

Structural connectivity, achieved through physical means, due to its more tangible form and its benefits for the preservation of biodiversity, has aroused interest and captivated various spheres working in conservation and planning. The great interest in connectivity corridors and their various uses has also generated debate and criticism about the concept of connectivity and the different means and scales of achieving it.

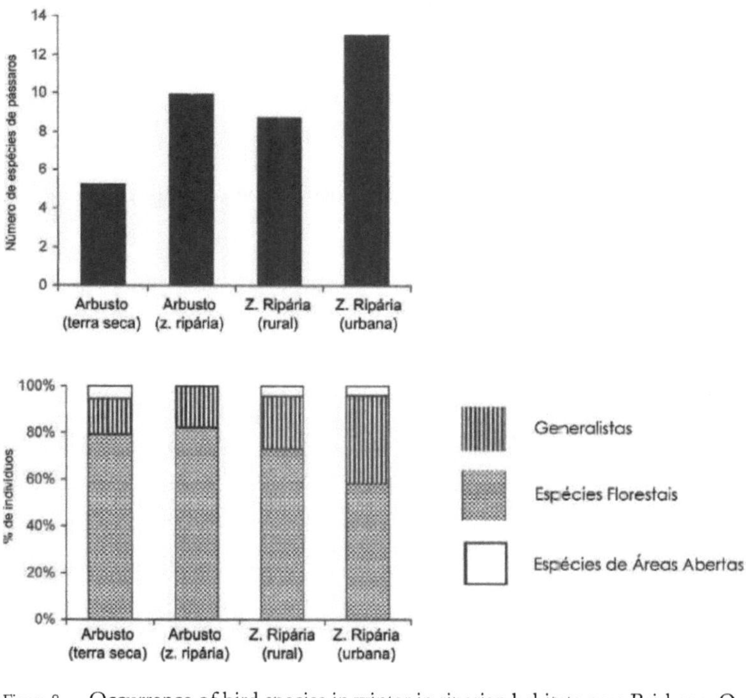

Figura 8: Occurrence of bird species in winter in riparian habitats near Brisbane, Queensland, Australia. (a) Comparison of species richness between dryland shrubs, riparian zone shrubs, strips of remnant riparian vegetation in rural areas and strips of riparian vegetation in urban areas, (b) Average percentage of individuals by different types of vegetation cover (Bennet, 2003).

Corridors and Connectivity

Ecological corridors are successful concepts, practical techniques with direct answers to real problems. "A visible solution to a visible problem" (Bennett, 2003). In this way, corridors have gained wide acceptance in the face of biodiversity conservation problems, but at the same time they have overshadowed other means of understanding and achieving connectivity.

Criticism and scepticism can be grouped into three general points (Noss, 1987; Simberloff & Cox, 1987; Harris & Gallagher, 1989; Harris & Scheck, 1991; Nicholls & Margules, 1991; Stolzenburg, 1991; Hobbs, 1992; Simberloff et al., 1992; Andrews, 1993; Bonner, 1994; Hess, 1994 apud Bennett, 2003):

- whether the scientific evidence is sufficient to demonstrate the potential conservation benefits of corridors;
- whether the possible negative effects of the corridors could outweigh their conservation value;
- whether corridors are cost-effective compared to other ways of conserving resources.

In addition to linear connectors, there are other ways of connecting habitats, such as stepping stones, mosaics of preserved areas; other forms of movement, such as migratory movements and daily movements; other ways of perceiving connectivity, such as ecological processes present within and between habitats.

There is little doubt that the movement of animals and plants, and the flow of wind, water, materials, nutrients, energy and biota between habitats is fundamental to the natural functioning of the ecosystem (Forman, 1995). The question that needs to be asked is:

What is the most efficient pattern in a given landscape that ensures ecological connectivity for species, communities and ecological processes? (Bennett, 2003). And furthermore, which ecological processes in the landscape show us the natural connectivity between habitats and ecosystems?

Movement

The concept of connectivity is intrinsically related to the ways in which species move and disperse. Daily or seasonal movements seeking to meet their biological and behavioural needs, or dispersal, seeking to increase their distribution area through movements from their place of origin (Noss, 1991).

Since different species have different ways of moving and dispersing, connectivity can be achieved, or obstructed, in different ways according to each species. Flying species are more mobile than terrestrial species. Even so, some bird species have psychological barriers to crossing aquatic environments (Noss, 1991). Plants dispersed by birds or wind can be quite mobile, while those dependent on ants will not have such a wide dispersal. A corridor for one species can be a barrier for another (Noss, 1991).
The particularities of each species generate physical, biological and behavioural differences, differences in movement and dispersal, etc., highlighting the importance of generating connectivity that covers and optimises a wide variety of habitats and a wide range of species.

Think of connectivity in a broader way, involving connections between species, communities, ecological processes and multiple spatial and temporal scales (Noss, 1991).

Multiple scales

Once the complexity of the natural world and all its biological processes is clear, so is the need to conserve biodiversity in all its integrity, covering the largest possible spatial and temporal scales.

Spatial Scale

The differences in movement and dispersal between species make obvious the different spatial scales present in the same landscape. Flying species tend to be more mobile than non-flying ones; carnivorous animals are generally less sedentary than herbivores; large terrestrial mammals have a totally different perception of scale size than terrestrial insects. The scalar spatial difference can occur even when analysing a single species, taking into account the different stages of its life: frogs begin life exclusively in aquatic environments, later moving on to occupy terrestrial environments as well (Bennett, 2003).

here is no general solution for maintaining connectivity for all species. Different qualities, scales and types of connections are needed for all species diversity and ecological processes.

Time scale

Landscapes, their ecosystems and habitats are not static, but change over time. These changes can result from natural processes or from disturbances caused by anthropogenic action. In both changes, there can be alterations in structure, ecological processes, natural dynamics and, consequently, the availability of resources for species (Bennett, 2003).

Dynamics within a population can also change over time. The number of individuals can increase with immigration rates, or decrease due to extinction. This fluctuation in the density of one species can influence others, interfering with the ecological processes that ensure connections between habitats.

Working with temporal and spatial scales makes it possible to visualise connectivity in its broadest sense, taking into account not only the structure of the landscape, but the natural dynamics and their connections.

Most of the discussions about connectivity that permeate wildlife planning and management research and projects are based on how to establish corridors. This neglects the important fact that connectivity is a natural element that already exists in the landscape.
Human action, in its various forms, can destroy or create barriers, making it difficult or impossible for connectivity to exist. What is desperately needed is to restore or maintain corridors or whatever the most appropriate form of connectivity is, not to "establish" it.

Thinking about connectivity in a broader and more comprehensive way provides a more complete understanding of the dynamics of nature, and gives us the certainty that there is still much to research and the incentive to continue researching.

CHAPTER 3

Sedimentation and island formation in the lagoon system

Thinking about connectivity through an island study and adopting a plant species as the main indicator and object of study, made the work understand a broader concept of connectivity.

The usual idea of islands as geographically isolated lands and the lack of mobility of individual plants can lead, at first, to the idea that there is no effective relationship between matter, flow and energy between the areas studied.

In this context, waterways connect the islands, moving sediment and nutrients from different sources and creating a third environment of great flow and exchange.

In a lagoon region full of well-developed mangroves, the abundance of nutrients and sediments further emphasises this idea of connectivity between the islands' forests. The composition of the sediments, the nature of the nutrients and the origin and formation of the region's bedrock are fundamental in characterising this connectivity.

Morro de São João, 120 metres high (Kutner, 1962) and made up of alkaline rocks (Freitas, 1947), is the only elevation on Cananéia Island. The rest of the island is made up of Quaternary sediments.

Suguio & Martin (1978) describe Ilha Comprida as a Quaternary barrier separated from the mainland by the Pequeno and Cananéia seas. It is a long, narrow island, 70 km long and no more than 5 km wide, with low altitudes of up to 7 metres. The only exception is Morrete, with 42 metres of alkaline rock.

A submerged crystalline outcrop appears to be the link between Morrete and Morro de São João in Cananéia, located in the southern part of their respective islands (Suguio & Martin, 1978).

The geological characterisation of the study area reveals the recent lagoon geomorphology of this region. This large sedimentary coastal plain dates back to the sedimentary deposits of the Cenozoic. Suguio and Petri (1973) characterised this stigraphy in four stages:

a) Sequence 1

118 metres thick;

sands and conglomeratic sands with clay levels;

probably covers the crystalline basement.

b) Sequence 2

silty clays rich in diatoms and poor in well-preserved foraminifera;

-indicative of a transitional environment.

c) Sequence 3

-7 to 12 metres thick;

-siltic reefs with abundant well-preserved foraminifera.

c) Sequence 4

- 20 to 30 metres thick;

. well-selected fine to very fine sand with good distribution;

-supplier of sediment for today's beaches.

The geomorphological development of this sedimentary plain is closely linked to the Quaternary transgressive phases. Sedimentological, biological and prehistoric evidence, identified through systematic mapping studies and radiocarbon dating of samples (Suguio & Tessler, 1992), proves the different marine palaeo-levels of the Brazilian coast. These transgressions were very important processes in the evolution of the Cananéia-Iguape lagoon region and other Brazilian coastal plains.

Among the different sea level scenarios, the highest marine palaeo-level of 120,000 years B.P. is known on the São Paulo coast as the Cananéia Transgression (Suguio & Martin, 1978) or, nationally, as the Penultimate Transgression (Bittencourt *et al.*, 1979).

The Cananéia Transgression played a predominant role in the sedimentary deposits of the Cananéia Formation (Suguio & Petri, 1973). The alignments of the beach ridges, when analysed by aerial photographs, also reveal the existence of a regressive phase. The basal clay part of the Cananéia Formation, first deposited in brackish waters and later in marine environments, demonstrates the trangressive phase (Suguio & Tessler, 1992).

In addition to the Cananéia Transgression, there were less extensive sandy deposits during the Santos Transgression and also well-developed mangrove deposits on the banks of the lagoon channels.

Suguio and Tessler (1992) describe the evolution of the coastal plain between Cardoso Island and the

Juréia in five stadiums:

First stage - at the apex of the Cananéia Transgression, the sea level probably reached the base of the Serra do Mar, depositing transitional and marine clays of the Cananéia Formation;

Second stage - already in the regressive phase, deposits of beach crystals covered the sandy deposits of the Cananéia Formation;

Third stage - around 18,000 years B.P., the sea level was 110 metres lower than it is today, which enabled the rivers of the recent coastal plain to erode the sandy sediments of the Cananéia Formation;

Fourth stage - the next transgressive event is characterised by a rapid rise in sea level which floods the areas eroded by the rivers, forming a large lagoon system;

Fifth stage - the return of the sea level to the current level generated Holocene deposits of prairie crystals.

The genesis and geomorphological evolution of the Cananéia-Iguape lagoon region is closely linked to the transgressive and regressive events of the Quaternary, the hydrodynamic processes acting on the coast and the atmospheric circulation of the southern part of South America (Suguio & Petri, 1973). This evolutionary process resulted in a semi-confined coastal plain, with typical siltation tendencies and sedimentary deposition features (Besnard, 1950).

Tidal movements and freshwater discharges have a strong influence on the geomorphological dynamics of the system, influencing the distribution of bottom sediments, guiding erosive phenomena and depositional features (Tessler, 1982; Tessler & Furtado, 1983).

The hydrodynamics of the lagoon channels in the Cananéia-Lguape system have already been studied and are well known. Sand banks, spurs and islands are frequent sedimentary features in the Cananéia and Cubatão Sea and in the Mar Pequeno (Cunha-Lignon, 2001). Tessler and Furtado (1983) cite the islands of Pai Matos, Boqueirão, Garça and Furadinho, all covered by mangroves, as sediment deposition points.

The Little Sea is the least meandering of the channels in the Cananéia-Lguape system. At its start, to the north of Cananéia Island, at the meeting of the Cubatão and Cananéia seas, it has a deeper, less meandering circulation channel with several sedimentary islands. In its northernmost part, from Pedra do Tombo to Barra de Icapara, the channel is more straight and has fewer sedimentary islands (Tessler & Furtado, 1983).

Figure 9. Schematic map of the lagoon channels of the Cananéia-Lguape system.

The Cubatão Sea, the narrowest in the system, has depths of more than 5 metres at the points of confluence with other channels (Cananéia Sea and Trapandé Bay). It is not very meandering and has few sedimentary islands (Tessler & Furtado, *op cit).*

The Cananéia Sea, the study area of this paper, is the most meandering and deepest of the system's channels. It contains various depositional features such as the various sedimentary islands (Pai Matos, Boqueirão, Garça, Furadinho, etc) and sand banks (Tessler & Furtado, *op cit).*

Tessler and Furtado (op *cit),* using nautical charts and aerial photographs, concluded that the depositional features of the Cananéia Sea have a preferential direction of growth, a trend linked to unidirectional flow originating from tidal currents. The elongation of the deposition features towards the Cananéia bar shows this predominance (Figures 10, 11 and 12).

The deposition features, as well as most of the hydrodynamic pattern of the system, are strongly and intimately linked to the currents generated by tidal oscillations (Tessler & Souza, 1998). The diffusion of tidal waves begins in Trapandé Bay, at the mouth south of the city of Cananéia (Miniussi, 1959).

During the flood, the tidal waves enter the system through the Cananéia bar, at the southern limit of Comprida Island, and divide into the channels that surround Cananéia Island, the Cubatão and Cananéia seas. At the northern limit of the island of Cananéia, the channels resume their single flow through the Pequeno Sea, towards the other tidal inlet at the mouth of the Icapara. The meeting of the opposing flows, both flood tides, is the hydrodynamic phenomenon known as "tumbling waters" (Tessler & Souza, 1998). At ebb tide, the flows reverse towards the connections with the ocean.

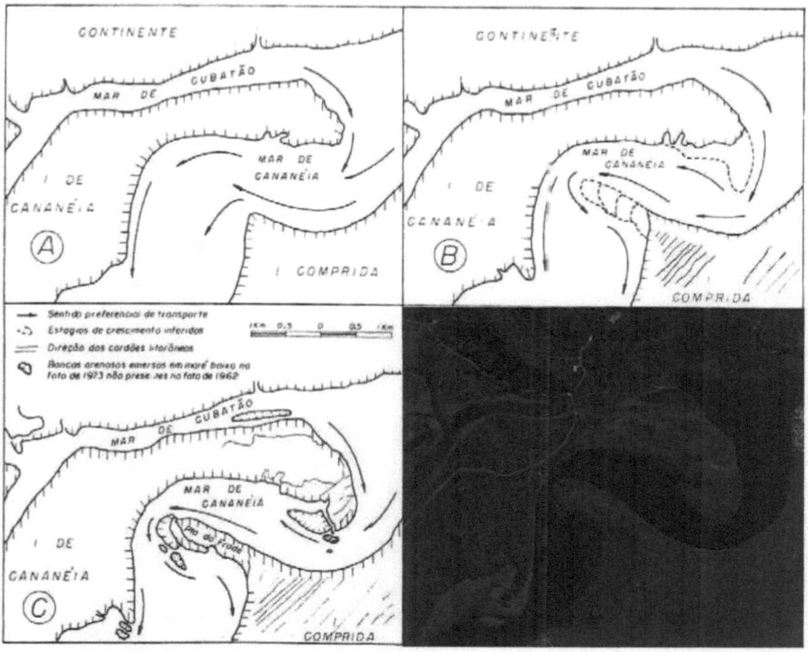

Figura 10: Schematic comparison of the evolution of Ponta do Frade: a) preliminary stage of the spur; b) intermediate phase; c) current configuration of Ponta do Frade (Tessler & Furtado, 1983). Satellite image amai (Google Earfh, 2010).

Tessler and Furtado (op.cit), based on nautical and bathymetric charts, aerial photographs and measuring stations, found that the maximum speed of tidal currents throughout the system was 1.2 m/s in the Cananéia Sea during the ebb period. The Cubatão and Iguape seas showed similar values for bottom and surface velocities, while in the Cananéia Sea the discrepancy was quite marked, especially at low tide.

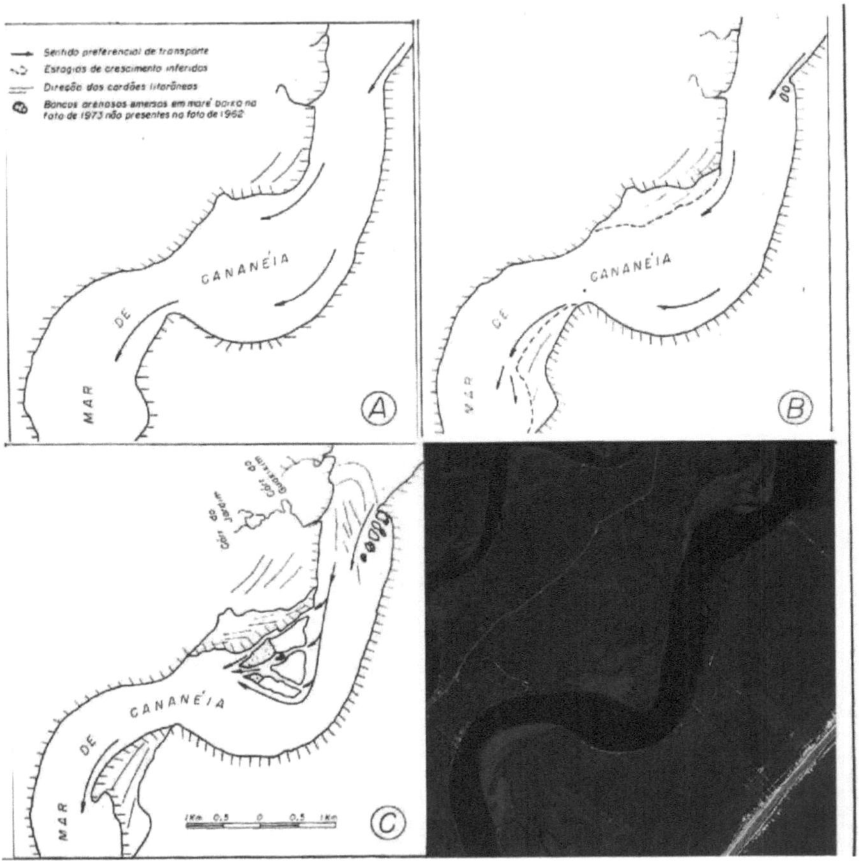

Figura 11: Schematic comparison of the evolution of Boqueirão Island: a) depositional features of Cananéia Island that will shape Boqueirão Island; b) sandbank incorporated by Cananéia Island; c) current configuration of Boqueirão Island (Tessler & Furtado, 1983). Current satellite image (Google Earth, 2010) - in detail, area of Figure Y .

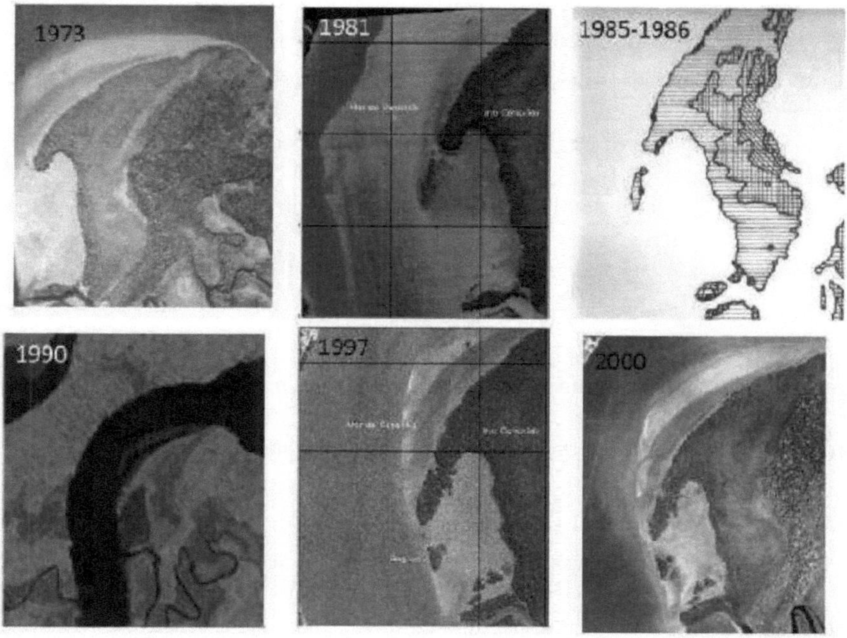

Figura 12: Aerial photographs and satellite images, at different temporal and spatial scales, of Baguaçu Island between 1973 and 2000 (Hertz, 1991; Cunha-Lignon, 2002).

Figura 13:

Despite the occurrence of currents with alternating directions at flood and ebb, and with opposite directions of movement in the surface and bottom layers, the higher tidal speeds at ebb, both bottom and surface, corroborate the preferential direction of the depositional features.

Field Analyses

Guided by the objectives to be achieved, the fieldwork was delimited in such a way as to optimise the whole process without compromising the expected results. The definition of the collection area was based on this assumption, together with the scope of the project, which shaped the work.

At first, it was thought that all the *Spartina* banks along the coast of Cananéia Island would be collected, but due to their large size, this would require more time than was planned for the dissertation. Furthermore, supported by an island biogeographical base, there was nothing wiser and more necessary than working on the relationship between islands, rather than singularities.

Just as an island was added, the mainland and the ocean were not included. The aim was to obtain results from the contact face between Cananéia Island and Ilha Comprida.

Starting from the USP Oceanographic Institute Base (southern limit), we travelled along the lagoon channel between the islands. We travelled along the Cananéia Sea, observing the shores as far as the village of Pedrinhas (northern limit). The geographical coordinates of the *Spartina* points were taken and then tabulated and mapped.

The Cananéia Sea is very sinuous, forming meanders and, consequently, shallow areas, sand banks and sedimentary islands colonised by mangroves (Tessler & Furtado, 1983). The species was found on the banks of the channel, river mouths and gamboas and on the outermost part of some islands.

Figura 13: The two red circles show the southern and northern limits of the shores of the Cananéia Sea, where the samples were taken.

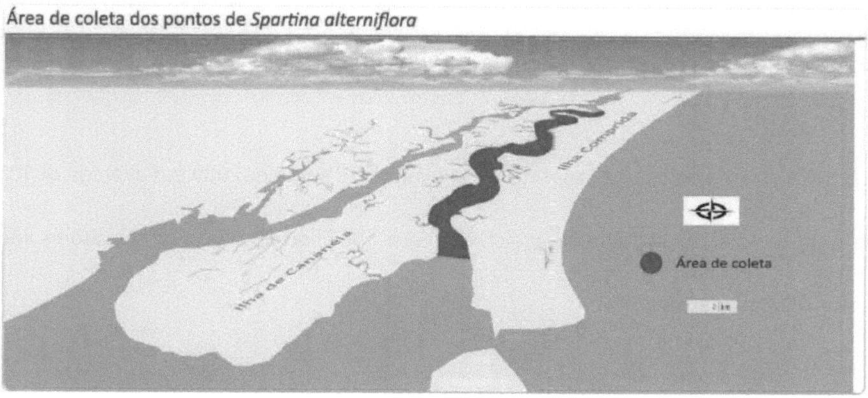

Figure 14: Infographic of the collection area for the Spartina alterniflora points.

Taking into account the characteristics of *Spartina alterniflora*, the field work was scheduled to cover the annual seasonal variations, representing our dry and rainy periods well.

The points were collected using a handheld Garmin e¯rex GPS device, using geographical coordinates (latitudes and longitudes), which were then converted to UTM.

All the points collected were listed in the field booklet, with the help of symbols identifying whether or not the samples were continuous. For areas less than 5 metres long, a dot was drawn next to the number that named the point. For more contiguous areas, an arrow connected the two numbers of the two points that limited the area. The dashed arrow indicates gaps in continuity.

Once the coordinates of the *Spartina* points were in hand, the latitude/longitude coordinates were converted to decimal degrees and tabulated. The data was collected and inserted according to the WGS84 *(World Geodetic System)* on a georeferenced image. It was also entered using the Universal Transverse Mercator system - UTM SAD-69 *(South America Datum),* making it easier to know distances in metres.

Field activities were carried out in spring 2008, winter and summer 2009. An average of 100 points were sampled in all the fields, which represent the *Spartina* banks on the shores and islands of the Cananéia Sea.

The banks with the presence of 5. *alterniflora* were classified according to their length, bordering the canal. The width of the banks was never measured. The mere presence of *Spartina altemiflora* on the banks of the canal, identifying its presence, would already be representative of the intention of the work.

The spatialisation of the banks provides information on the possible sites of sedimentation and perhaps colonisation and expansion of the mangrove forests.

Results

The maps produced from the points collected record the presence, location and extent of the *Spartina altemiflora* banks. The data was collected in spring and winter, which characterises the height of the grass's growth and its most dormant period.

The collection on both shores of the Cananéia Sea sought to show a possible relationship between the islands of Cananéia and Comprida. This relationship can be realised through the flow of energy, nutrients and, as far as this research is concerned, sediment transport.

The maps show the places in the Cananéia Sea where there is the greatest potential for *Spartina* colonisation. *The* low-energy meanders, where sediment accumulates, proved to be favourable sites for the presence of 5. *altemiflora* banks

This characterisation of *Spartina*'s geographical distribution was extremely important, as it provided the basis for the grass's spatial behaviour.

CHAPTER 4

Conclusion

Spartina alterniflora and sedimentation

In a semi-confined environment such as the Cananéia-Iguape lagoon system, which is prone to siltation and consequent sedimentation, the interaction between substrate and vegetation is extremely important, as it creates the conditions for future spatial organisation. At first, pioneer species colonise the exposed sediment deposits and then facilitate the colonisation of other species (Connel and Slatyer, 1977 *apud* Cunha-Lignon, 2009). This happens in the lagoon channels of Cananéia, where the sand banks are colonised by *Spartina alterniflora,* holding and preparing the substrate for seedlings of tree species from the region's mangroves.

As early as 1940, Davis identified the presence of *Spartina alterniflora* in pioneer communities and its fundamental role in sediment stability and the establishment of young *Rhizophora* seedlings.

A case study by Cunha-Lignon (2009) analysed the spatio-temporal variation of mangrove forests. By analysing satellite images, aerial photographs and monitoring fixed plots, the structure, zonation and succession of mangrove forests were analysed. A comparison of photographs from the southern coast of the state of São Paulo showed a positive relationship between increased areas of sediment deposition and increased mangrove area. The sediment banks colonised by mangrove forests were first occupied by *Spartina alterniflora* (Figure 15).

The highest density of seedlings and young mangrove individuals colonise the 5. *alterniflora* strip, *with* little colonisation on the sediment bank. The colonisation of the seedlings follows the area of expansion of the grass (Cunha-Lignon, 2005) (Figure 16).

The presence of *Spartina alterniflora* is of fundamental importance in fixing sediments and establishing propagules. Higher densities of seedlings and young mangrove individuals are found in areas colonised by the grass, while on exposed sediment banks few propagules are established (Cunha-Lignon, 2006) (Photo 1).

Spartina alterniflora on sediment banks in mangrove areas enables, prepares and facilitates colonisation by mangrove seedlings, promoting the expansion of the mangrove forest. As this expansion takes place, the mangrove forests replace the strips previously occupied by the grass.

The presence of sediment banks occupied by *Spartina alterniflora* on the outer fringes of the mangroves, on the facies in contact with the channels, shows the potential for the establishment and expansion of mangrove forests.

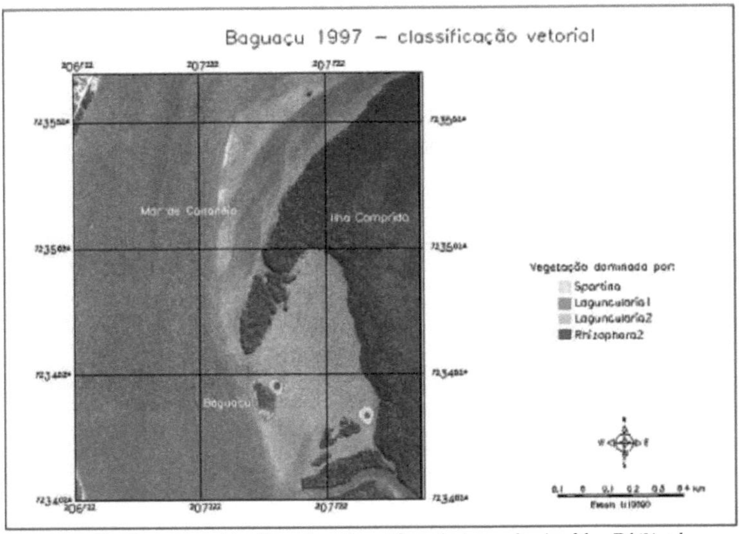

Figura 15. Banks covered by Spartina alterniflora being colonised by *Rhi%ophora mangle* (Cunha-Lignon, 2005).

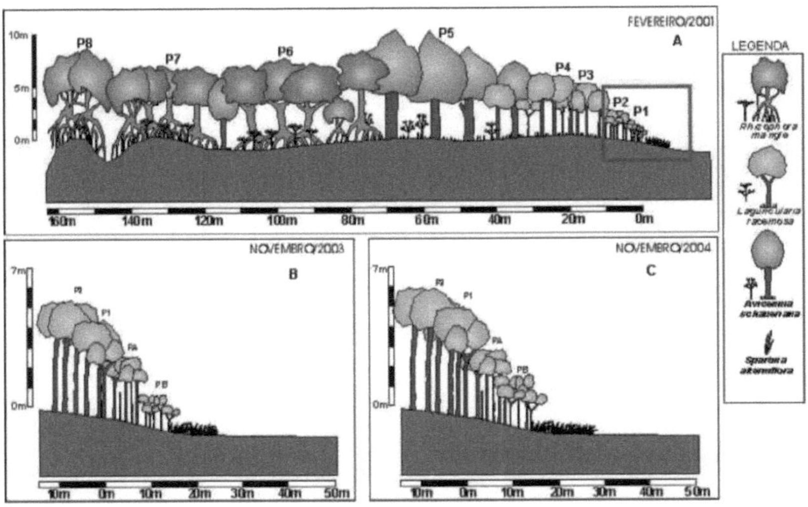

Figura 16. Structural development of the mangrove forest on Baguaçu Island: a) plots studied by Cunha-Lignon, 2001; b) monitoring in 2003; c) forest structure in 2004 (Cunha-Lignon, 2005).

Photo 2. *Spartinu :lta)niflora* bank being colonised by *Rhi^opUorb mangleU* a) June, 2001; b) December, 2003; c) November 2004 (Cunha-Lignon, 2005).

CHAPTER 5

Final considerations

The role and importance of the *Spartina alterniflora* grass to the mangrove ecosystem in which it is found in the Cananéia-Lguape lagoon system is clear.

Identifying sites with a tendency for Spartina to grow, understanding the characteristics that generate this tendency, understanding the behaviour of the grass during and after its establishment and the possibilities it generates for the colonisation of mangrove species; this data provides us with knowledge of flow (currents) and matter (sediments).

The characterisation of the species' geographical distribution made it possible to analyse the relationship between low energy, sedimentation and *Spartina*'s pioneering spirit. A long-term study could show the stages of colonisation of mangrove forests after the establishment and permanence of this relationship.

The probable development of mangroves on these Spartina banks, as other studies have shown, would further contribute to their role of protecting and supplying nutrients to coastal environments.

As for connectivity, the main aim of this work was to promote and instigate knowledge of its various spheres of action. The different scales at which connectivity operates show that its importance goes beyond the tangible, awakening us to an understanding of the system as a whole.

Obtaining data showing the origin and probable destination of sediments and nutrients would require laboratory analyses that are beyond the scope of this work. Such data would certainly enrich the discussion.

All the information gathered and analysed, mainly around the triad *of Spartina alterniflora*, connectivity and sedimentation, meant that this research achieved its main objective.

Bibliography

Ab'Saber, A. N. (2003) *The domains of nature in Brazil: landscape potential.* São Paulo: Ateliê Editorial.

Adaime, R. R. (1977) Study of the stationary variation of the standing crop and the restocking of a bank of 5. *alterniflora* Loiseleur, 1807, in the Cananéia estuarine-lagoon complex. Master's Thesis, University of São Paulo, São Paulo, Brazil.

Adams, D. A. (1963) Factors influencing vascular plant zonation in North Carolina salt marshes. Ecology,Vol.44, No.3, pp. 445-456.

Alves, L. F. (2004) Effect of connectivity on the abundance and richness of seedlings and juveniles of tree species in fragmented landscapes. Technical Report - Post-Doctoral Fellowship/FAPESP (Process No. 00/03945-6) linked to the BIOTA/FAPESP Programme: Biodiversity Conservation in Fragmented Landscapes on the São Paulo Atlantic Plateau

(Process No. 99/05123-4). Sub-Project 7.

Alves, L. F. and J. P. Metzger, J. P. (2002) Fragmentation effects on tree seedling density in a tropical agricultural landscape of southeastern Brazil. In: 45th Symposium of the International Association for Vegetation Science, Porto Alegre, Brazil, p. 102

Bennet, A.F. (2003) Linkages in the landscape: the role of corridors and connectivity in wildlife conservation. IUCN, Gland, Switzerland and Cambridge, UK.

Bérgamo, A.L. (2000) Characteristics of hydrography, circulation and salt transport: Barra de Cananéia, South Cananéia Sea and Trapandé Bay. Master's thesis, University of São Paulo, Oceanographic Institute, USP, 21 Op.

Bertrand, G. (1971) *Paisagem e Geografia Física Global: esboço metodológico.* University of São Paulo: Institute of Geography.

Besnard, W. (1950) General considerations about the Cananéia-Iguape lagoon region. Bolm. Inst.
Paul. Oceanogr., 1 (1): 3-28.

Besnard, W. (1950a) Considerações gerais sobre em torno da região lagunar de Cananéia-Iguape I. Boi. Inst. Paul. Oceanogr., São Paulo, 1 (1): 9-26.

Besnard, W. (1950b) Considerações gerais sobre em torno da região lagunar de Cananéia-

Iguape II. Ox. Inst. Paul. Oceanogr., São Paulo, 1 (2): 3-28.

Bianchini Jr., I. (1982) Contribution to the study of the decomposition of aquatic plants. Master's Thesis, Federal University of São Carlos, São Carlos, Brazil.

Bianchini Jr., I. (1984) Studies of the humification processes of *Nymphoides indica* (L.) O. Kuntze.
PhD Thesis, Federal University of São Carlos, São Carlos, Brazil.

Bittencourt, A. C. S. P.; Vilas-Boas, G. S. & Flexor, J. M. (1979) Quaternary marine formations of the coast of the State of Bahia, Brazil. *In:* International Symposium on coastal evolution in the quaternary, São Paulo, 1978. Proceedings. São Paulo, 1978, p. 232-253.

BRASIL. (2003) *Fisheries Statistics 2001 - Major Regions and Units of Brazil.* Tamandaré - PE: Ministry of the Environment, Brazilian Institute of the Environment and Renewable Natural Resources.

BRASIL. (2004) *Fisheries Statistics 2002 - Major Regions and Units of Brazil.* Tamandaré - PE: Ministry of the Environment, Brazilian Institute of the Environment and Renewable Natural Resources.

BRASIL, (undated) *Fisheries Statistics 2000 - Major Regions and Units of Brazil.* Ministry of the Environment, Brazilian Institute of the Environment and Renewable Natural Resources.

Brown, J.H. & Kodric-Brown, A. (1977) Rates in insular biogeography: effect of immigration on extinction. Ecology, Vol.58, No.2, pp. 445-449.

Bunn, A.G., Urban, D.L. & Keitt,T.H. (2000) Landscape connectivity: a conservation application of graph theory. Journal of Environmental Management, Vol.59, pp. 265-278.

Cammen, L. M.; Rublee, P.; Hobbie, J. (1978) Bacteria in a North Carolina salt marsh: standing crop and importance in the decomposition of *Spartina alterniflora.* Univ. of North Carolina Sea Grant Publ. UNC-SG 74-75.

Cammen, L.M. (1980) The significance of microbial carbon in the nutrition of deposit feeding polychaete Nereis succinea. Marine Biology, Vol.61, pp. 9-20.

Chapman, V. J. (1974) Salt marshes and deserts of the world. 2nd Ed. J. Cramer Verlag, Leutershausen, Germany.

Christiansen, T.; Wiberg, P. L.; Milligan, T. G. (2000) Flow and sediment transport on a tidal salt marsh surface. Estu a ri ne, Coastal and Shelf Science, Vol.50, pp. 315-331.

Cody, M.L. & Diamond, J.M. (ed) (1982) Ecology and evolution of communities.The Belknap Press of Harvard University Press, England.

Costa, C. S. B.; Davy, A. J. (1992) Coastal saltmarsh communities of Latin America. *In* Seeliger, U. (ed) Coastal plant communities of Latin America. Department of Oceanography, University of Rio Grande, Rio Grande, Brazil.

Cowen, R.K. et al. (2000) Connectivity of marine populations: open or closed? Science Vol.287, pp. 857-859.

Craft, C; Megonigal, P.; Broome, S.; Stevenson, J.; Freese, R.; Cornell, J.; Zheng, L.; Sacco, J. The pace of ecosystem development of constructed *Spartina alterniflora* marshes. Ecological Applications, Vol.13, No.5, pp. 1417-1432.

Cunha-Lignon, M. (2001) Mangrove dynamics in the Cananéia-Iguape System, State of São Paulo - Brazil. Master's thesis, Oceanographic Institute, University of São Paulo.

Cunha-Lignon, M. (2001) Mangrove dynamics in the Cananéia-Iguape System, State of São Paulo - Brazil. Master's thesis, University of São Paulo, Oceanographic Institute of São Paulo.

Darnell, R. M. (1967a) The organic detritus problem. *In* Lauff, G. H. (ed) Estuaries. American Association for the advancement of Science, No.83.

Darnell, R. M. (1967b) Organic detritus in relation to the estuarine ecosystem. *In* Lauff, G. H. (ed)

Estuaries. American Association for the advancement of Science, No.83.

Davis, J. H. (1940) The ecology and geologic role of mangroves in Florida. Carnegie Institute Washington Publication, n. 517, p. 303-412.

Diamond, J.M. (1975) The island dilemma: lessons of modern biogeographic studies for the design of natural reserves. Biological Conservation Vol. 7, no. 2, pp. 129-146.

Diegues, A. C. S. (2000). *Ethnoconservation: new directions for nature conservation.* São Paulo: Nupaub/USP.

Diegues, A. C. S. (2001). *Human ecology and planning in coastal areas.* São Paulo: Nupaub/ USP.

Dunning, J.B.; Danielson, BJ. & Pulliam, H.R. (1992) Ecological precesses that affect populations in complex landscapes. Oikos, Vol.65, No.1, pp. 169-175. Nordic Society Oikos.

Forman, R. T. T. & Godron, M. (1986) *Landscape ecology.* New York: John Wiley & Sons Inc.

Forman, R.T.T. (1995) Land mosaics: the ecology of landscapes and regions. Cambridge University
Press, UK.

Freitas, R. O. de (1947) Alkaline eruptions of Cananéia, State of São Paulo. Boi Fac. Fil. Ciên. Letras da Univ. de São Paulo, no. 91, Geol. no. 4, São Paulo.

Fritz, R. & Merriam, G. (1992) Fencerow habitats for plants moving between farmland forests.
Biological Conservation 64, pp. 141-148. Elsevier Science Publishers Ltd, England.

Goodman, P. J. (1958) *Spartina Schreb.* The Journal of Ecology, Vol.57, No.1, pp. 285-287.

Gosselink, J. G.; Kirby, C. J. (1974) Decomposition of Salt Marsh Grass, Spartina alterniflora Loisel.
Limnology and Oceanography, Vol.19, No.5, pp. 825-832.

Guerra, A. J. T.; CUNHA, S.B. da (Org) (1998) *Geomorphology of Brazil.* Rio de Janeiro: Bertrand
Brazil.

Haines, E. B. (1977) The origins of detritus in Georgia salt marsh estuaries. Oikos, 29, pp. 254-260.

Helliwell, D.R. (1975) The distribution of woodland plant species in some shroshire hedgerows.
Biol. Conserv. 7, pp. 61-72. Applied Science Publishers Ltd, England.

Henein, K. & Merriam, G. (1990) The elements of connectivity where corridor quality is variable.
Landscape Ecology, Vol.4, Nos. 2/3, pp. 157-170.

Hobbs, RJ. & Saunders, D.A. (ed) (1992) Reintegrating fragmented landscapes: towards sustainable production and nature conservation. New York: Springer-Verlag.

Holt, R.D., Robinson, G.R. & Gaines, M.S. (1995) Vegetation dynamics in an experimentally fragmented landscape. Ecology, Vol.76, No.5, pp. 1610-1624.

Kareiva, R & Wennergren, U. (1995) Connecting landscape patterns to ecosystem and population processes. Nature, Volume 373, Issue 6512, pp. 299-302.

Karniol-Marquez, M. da R. (2007) Morphodynamics of a segment of Ilha Comprida beach, south coast of the State of São Paulo. Master's thesis. Ocenographic Institute, University of São Paulo.

Kutner, A. S. (1962) Granulometria dos sedimentos de fundo da região de Cananéia, São Paulo. Boi. Soc. Bras. Geol. vol. 11, no. 2, p. 41-45, São Paulo.

Leonard, L. A.; Wren, P. A.; Beavers, R. L. (2002) Flow dynamics and sedimentation in *Spartin alterniflora* and *Phragmites australis* marshes of the Chesapeake Bay. Wetlands, Vol.22, No.2, pp. 415-424.

MacArthur, R.H. & Wilson, E.O. (1967) The theory of island biogeography. Princeton, N.J.: Princeton University Press.

Mann, K. H. (1975) Relationship between morphometry and biological functioning in three Coastal inlets of Nova Scotia. Estuarine Research, Vol. 1, pp. 634-644

May, M. S. (1974) Probable agents for the formation of detritus from the halophyte, Spartina alterniflora. *In* Reimold, R. J. (ed) Ecology of Halophytes. University of Georgia Marine Insitute,

Sapelo Island, Georgia, USA.

McKee, K. L. (1988) The relation of smooth cordgrass *(Spartina alterniflora)* to tidal datums: a review. Estuaries, Vol.11, No.3, pp. 143-151.

Mendelssohn, I. A. (1979) The influence of nitrogen level, form, and application method on the growth response of Spartina alterniflora in North Carolina. Estuaries, Vol.2, No.2, pp. 106-112.

Mendonça, F. (2001) *Geografia socioambiental.* São Paulo: Terra Livre, 16:139-158.

Mesquita, A.L. de & Harari, J. (1983) Tides and gauges of Cananéia and Ubatuba - Brazil (Lat 24°).

Int. Int. Inst. Oceanogr. Univ. S. Paulo, (11): 1-14.

Metzger, J. P. (2001) *What is landscape ecology?* São Paulo: Department of Ecology, Institute of Biosciences/USP. Available at: <http://www.biotaneotropica.org.br>. Accessed on: 23 May 2007.

Metzger, J. P. (2003) Biodiversity Conservation in Fragmented Landscapes on the Atlantic Plateau of São Paulo. III FAPESP Annual Report (BIOTA/FAPESP Programme, Process No. 99/05123-4)

Metzger, J.P. & Décamps, H. (1997) The structural connectivity threshold: an hypothesis in conservation biology at the landscape scale. Acta Oecologica Vol.18 (1), pp. 1 -12.

Miniussi, I. C. (1959) Tidal wave propagation around Cananéia Island. Contr. Univ. São Paulo, Oceanogr. fís. series, (2): 1 -8.

Miranda, E. E. de; Coutinho, A. C. (Coord.) (2004) *Brasil Visto do Espaço.* Campinas: Embrapa Satellite Monitoring. Available at: <http://www.cdbrasil.cnpm.embrapa.br>. Accessed on: 02 July 2005.

Mobberley, D. G. (1956) Taxonomy and distribution of the genus *Spartina.* Iowa State College Journal of Science, Vol.30, No.4, pp. 471-574.

Moraes, A. C. R. (1999) *Contribuições para a gestão da zona costeira do Brasil: elementos para uma geografia do litoral brasileiro.* São Paulo: Hucitec; Edusp.

Nixon, S. W.; Oviatt, C. A. (1973a) Analysis of local variation in the standing crop of *Spartina alterniflora.* Botanica Marina, Vol.16, pp. 103-109.

Nixon, S. W.; Oviatt, C. A. (1973b) Ecology of a New England Salt Marsh. Ecological Monographs, Vol.43, No.4, pp. 463-498.

Noss, R.F. (1991) Landscape connectivity: different functions at different scales. pp.27-39 in Landscape linkages and biodiversity. (Ed. W. E. Hudson). (Island Press: Washington DC).

Noss, R.F. & Larry, D.H. (1986) Nodes, networks and MUMs: preserving diversity at all scales. Environmental Management. Vol.10, No.3, pp. 299-309. Springer-Verlag New York Inc.

Od u m, E. P. (1965) *Ecologia: estructura* y *funcion de la natureza. Modern principles of energy flows and biogeochemical cycles.* Mexico: Continental.

Odum, E. P.; de la Cruz, A. (1967) Particulate detritus in a Georgia salt marsh-estuarine ecosystem.

In Lauff, G. H. (ed) Estuaries. American Association for the advancement of Science, No.83.

Panitz, C. M. N. (1992) Ecological aspects of a saltmarsh ecosystem in Santa Catarina Island, Brazil. *In* Seeliger, U. (ed) Coastal plant communities of Latin America. Department of Oceanography, University of Rio Grande, Rio Grande, Brazil.

Pereira, O. M. *et al.* (2001) *Evaluation of the Crassostrea brasiliana oyster stock in rivers and lagoons of the Cananéia estuarine-lagoon region.* São Paulo: B. Inst. Pesca, 27(1): 85-95.

Pereira, O. M. et *al.* (2000) *Stock assessment of the oyster Crassostrea brasiliana (Lamarck, 1819) in the mangrove swamp of the Cananéia estuarine-lagoon region.* São Paulo: B. Inst. Pesca, 26(1): 49-62.

Pereira, O. M. *et al.* (1985) *Hydrography of the Cananéia estuarine-lagoon complex: salinity and temperature (1973 to 1980).* São Paulo: B. Inst. Pesca, 12(3): 109-21.

Pereira, O. M. *et al.* (1986) *Hydrography of the Cananéia estuarine-lagoon complex: influence of the tide.* São Paulo: B. Inst. de Pesca, 13(1): 51-70.

Preston, RW. (1962) The canonical distribution of commonness and rarity: part II. Ecology, Vol.43, No.3, pp.410-432. Ecological Society of America.

Redfield, A. C. (1965) Ontogeny of a salt marsh estuary. Science, Vol.147, No.3653, pp. 50-54.

Reed, D. J. (1989) Patterns of sediment deposition in subsiding coastal salt marshes,Terrebonne Bay, Louisiana:the role ofwinter storms. Estuaries, Vol.12, No.4, pp. 222-227.

Reitz, P. R. (1961) The vegetation of the maritime zone of Santa Catarina. Sellowia, Vol.13, No.13, pp. 17- 115.

Ross, J.F.S. (1997) Geomorphological map of the State of São Paulo. Geomorphology Laboratory, Department of Geography. Faculty of Philosophy, Literature and Human Sciences, University of São Paulo.

Sakamoto, L. *Water Farms.* Available at: <www.reporterbrasil.com.br/reportagens/pesca/pesca.pdf>. Accessed on: 30 June 2005.

Scott, P. C. (1986) *Considerations on the use of Sepetiba Bay, RJ for Mariculture supported by a Geographic Information System.* Institute of Biological and Environmental Sciences, Santa Úrsula University, RJ.

Silva, J.F. (1989) Climatological data for Cananéia and Ubatuba (State of São Paulo). Bolm. Climatol. Inst. Oceanogr., São Paulo, (6): 1-21.

Silveira, J. D. (1970) *Morphology of the Coast.* In: AROLDO, A. *Brasil, a terra e o homem.* São Paulo: Nacional.

Soctchava, V. B. (1978) *Por uma teoria de classificação de geossistemas de vida terrestre.* University of São Paulo: Institute of Geography.

Spinelli, F.P. (2003) The alkaline rocks of Cananéia, south coast of São Paulo State: petrographic, mineralogical and geochemical characteristics. Master's thesis, University of São Paulo, Institute of Geosciences, 134p.

Squiers, E. R.; Good, R. E. (1974) Seasonal changes in the productivity, caloric content, and Chemical composition of a population of salt-marsh cord-grass *(Spartina alterniflora).* Chesapeake Science, Vol.15, No.2, pp. 63-71.

Suguio, K & Martin, L. (1978) Quaternary marine formations of the São Paulo and southern Rio de Janeiro coasts. In: International Symposium on Coastal Evolution in the Quaternary. Special publication. Brazilian Geological Society. Institute of Geosciences, University of São Paulo. 55p.

Suguio, K. & Martin, L. (1978) Quaternary marine formations of the States of São Paulo and Southern Rio de Janeiro. In: International Symposium on Coastal evolution in the quaternary. São Paulo, 1978. Special Publication, no. 1,55p.

Suguio, K. & Martin, L. (1986) Quaternary marine formations of the State of São Paulo and Southern Rio de Janeiro. Intern. Symposium on Coastal Evolution in the Quaternary, Special Publication, São Paulo, Soc. Bras. Geologia, Vol. 1,55p.

Suguio, K. & Petri, S. (1973) Stratigraphy of the Iguape-Cananéia lagoonal region sedimentary deposits. São Paulo, Brazil. Part I: Field observations and grain size analysis. Bolm IG, Inst. Geociên., Universidade de São Paulo, 4:1 -20.

Suguio, K. & Tessler, L. A. P. (1992) Quaternary deposits of the Cananéia-Iguape coastal plain (SP). Tour itinerary of the 37th Brazilian Congress of Geology, Vol 1, 31 p.

Suguio, K. & Tessler, M. G. (1992) Quaternary deposits of the Cananéia-Iguape coastal plain (SP). Special Publication of the Oceanographic Institute of São Paulo, n.9, p. 1-33.

Taylor, P.D.; Fahrig, L.; Henein, K. & Merriam, G. (1993) Connectivity is a vital element of landscape structure. Oikos, Vol.68, No.3, pp. 571 -573. Nordic Society Oikos.

Teal, J. M. (1962) Energy Flow in the Salt Marsh Ecosystem of Georgia. Ecology, Vol.43, No.4, pp. 614-62

Tessler, M. G. & Furtado, V. V. (1983) Sedimentation dynamics of siltation features in the Cananéia-Iguape lagoon region, São Paulo State. Bolm. Inst. Oceanogr., São Paulo, 32 (2): 117-124.

Tessler, M. G. (1982). Current sedimentation in the Cananéia-Iguape lagoon region, State of São Paulo. Master's thesis, University of São Paulo, Institute of Geosciences, v. 1 (text) 110p. and v. 2 (maps and figures) 170p.

Tessler, M. G. (1994) *Geology of the continental and coastal shelf. Meteorology. Ocean and coastal environmental diagnosis of the south and southeast regions of Brazil.* São Paulo: Fundespa - Iousp - Petrobras.

Tessler, M. G. & Souza, L. A. P. (1998) Sedimentary dynamics and sedimentary features identified on the bottom surface of the Cananéia-Iguape system, SP. Rev. Bras. de Oceanogr., 46 (1): 69-83.

Tessler, M. G. & Souza, L. A.P. de (1998) Sedimentary dynamics and sedimentary features identified on the bottom surface of the Cananéia-Iguape system, SP. Rev. bras. oceanogr., 46 (1)

Tischendorf, L. & Fahrig, L. (2000) How should we measure landscape connectivity? Landscape Ecology, Vol. 15, pp. 633-641.

Tomlinson, P. B. (1986) The botany of mangroves. Cambridge University Press, Cambridge,

England.

Turner, R. E. (1976) Geographic variations in sat marsh macrophyte production: a review. Contributions in Marine Science, Vol.20, pp. 47-68.

Urban, D. & Keitt, T. (2001) Landscape connectivity: a graph-theoretic perspective. Ecology, Vol.82, No.5, pp. 1205-1218.

Valenti, W. C. (2000) *Aquaculture in Brazil: bases for sustainable development.* Ministry of Science and Technology.

Valiela, I. (1974) Nutrient limitation in a salt marsh vegetation. *In* Reimold, R. J. (ed) Ecology of Halophytes. University of Georgia Marine Insitute, Sapelo Island, Georgia, USA.

Vanucci, M. (2002) *Mangroves and us: a synthesis of perceptions.* São Paulo: Edusp.

Vrieling, A. & Sanden, J. van der. (2000) *Satellite remote sensing for monitoring coastline dynamics ofthe Canadian beaufortsea coast.* Canada Centre for Remote Sensing.

Whittaker, R. J. (1998) *Island biogeography: ecology, evolution, and conservation.* New York: Oxford University Press.

Wiegert, R. G. (1977) Ecological processes characteristic of Coastal Spartina marshes ofthe south- eastern U.S.A. In Jefferies, R. L. (ed) Ecological processes in Coastal environments.

Blackwell Scientific Publications.

Wilson, E.O. & Willis, E.O. (1975) Applied Biogeography. In: Cody, M. L. & Diamond, J. M. (ed) Ecology and evolution of communities. Belknap Press of Harvard University Press: Cambridge, Massachusetts and London: England.

Zalewski, M. (2004) Do smaller islands host younger populations? A case study on metapopulations ofthree carabid species. Journal of Biogeography, Vol.31, pp. 1139-1148.

Zollner P.A. & Lima, S.L. (1999) Strategies for landscape-level interpatch movements. Ecology, Vol.80, No.3, pp. 1019-1030.

Zunino, M. & Zullini, A. (2003) Biogeography: the spat al dimension of evolution. Mexico: Fondo de Cultura Económica.

MIX
Papier aus verantwortungsvollen Quellen
Paper from responsible sources
FSC® C105338

Printed by Books on Demand GmbH, Norderstedt / Germany